God's Blessing The Human Body

PALMETTO
PUBLISHING
Charleston, SC
www.PalmettoPublishing.com

God's Blessing The Human Body
Copyright © 2024 by Edythe J. Rice

First Edition

Hardcover ISBN: 979-8-8229-3850-2
Paperback ISBN: 979-8-8229-3851-9

God's Blessing
The Human Body

EDYTHE J. RICE

This book is dedicated to my Mom and Stepdad,
Mrs. Hedy Henderson and Elder Chester Henderson.

These beloved people led me to Christ. I will forever be
indebted and grateful to them. They will forever be in my
thoughts and prayers, your everloving daughter.

THE HUMAN BODY

GOD IS THE CREATOR OF HEAVEN AND EARTH. All things made is made by him there is no greater god.

There is a lot of talk today about AI; artificial intelligence, but i say, GI; God's intelligence is greater. No other intelligence can trump god's. This statement is mine only.

God created the sun, the moon, the stars, the trees, the seas, lakes, oceans, the seeds, the grains, the stones for our very existence. God said, let's make man in our image.

God is a masterful god that he made our bodies so wonderful that it would serve us well all the days of or lives.

I want people to know how wonderful our bodies are. . Compare it, if you will to a sophisticated, complex machine. It can multitask, cleanse, regenerate, heal, to name a few of the functions.

You can even donate an organ in order to save someone's life. That is truly amazing.

I won't even attempt to list everything that each organ or tissue functions, but i truly want you to be acquainted with your body and what it does. Please take care of it. It will serve you well.

It is often asked when you write a book, who do you want to target. I want to target everyone, but particularly, the youth. The youth because the earlier you are aware of your body and what it does, you are apt to make better judgment in your lifestyle.

Taking care of your body early on can and will maximize your days on this beautiful earth.

It is my fervent prayer that after reading this book, you have a greater appreciation about your unique body.

With the greater knowledge about your body, you can and will appreciate your body.

God gave you a masterpiece. Your machine is like no other, a one of a kind.

Take care of your body and it will serve you well. You can live a long, healthy life. It's up to you. With that being said, let's get started.

THE BRAIN

THE BRAIN I UNDERSTAND IS A SMALL ORGAN, but it carries a mighty punch. The brain is like the president, ceo of your body. It weighs about three pounds, so small, yet probably the most important of the organs. The brain is fully formed at age 25. The brain controls and tells the other organs when and how to perform.

The brain performs so many functions that it is hard to imagine how much it does and how much we are blessed to have this organ.

The brain also controls our thoughts, our feelings, our emotions. It helps with our memory, our vision. The brain i understand is so soft and spongy with many grooves and crevices. The brain resides, lives in the skull. The brain also monitors our blood sugar levels which is important to diabetics.

The brain is a highly functioning organ, complex and truly marvelous.

Did you know that the brain is 75 to 80 percent water. The brain is the most complex of our organs. The brain really does multitask. Some part of the brain regulates our sexual function. It really is functioning in almost every aspect of our bodies. This organ really performs.

The brain is so complex. One of its function is to recognize faces, identify danger, and movements.

The brain stores so much memory for us. Some memory is short and some long term.

You are fearfully and wonderfully made.

You can keep your brain active by exercising it with reading a book, by writing down your thoughts, goals and desires. Also, helpful would be to write down your five year, ten year plan. You can always go back and see what has come to fruition. You can literally see your progress.

Keep your brain active by always thinking, your goals, your dreams can always change, but write down your new goals.

I think we lose so much in life by not writing things down. It makes us think.

Thinking and feeling is so important for the brain. The brain is always working. It never fails.

Thinking is so amazing as you are always learning. Life is never dull, but is an open door to new experiences. Keep your brain alive. It is truly wonderful.

The brain is so amazing that i totally forgot about an incident that happened when I was sixteen.

I was dating this young man, he was eighteen or nineteen. We would go to the movies, go to get ice cream and go to visit a mutual friend. One day after coming home from the movies, he asked me if i would do him a favor at the appropriate time and i said yes. After several more dates and coming home from this date, he drove me home, but the route he took was not the usual one.

He kept driving to where i didn't know. He stopped the car and said, it's time.

A ton of bricks hit my head and immediately I knew what the favor was and that I had consented to grant it.

My brain realized what i said i would do, but i didn't know that what I had agreed to was to go to bed with him.

I thought the favor I agreed to was perhaps to pay for our date or something like that. He was always a gentleman, so I had no reason to think unfavorably of him.

Talking about being naive, reality showed up.

After being confronted about this promise I said i would do, I then thought about what could happen to me. I could be raped, beaten or even killed.

I don't know what happend, but my body started shaking. I shook so much that it apparently scared the young man that he rushed me home without saying a word or looking at me. I didn't hear from him for several months. When I did hear from him, his question to me was why was I violently shaking at that time. I told him I didn't know why or what i would have done.

The shaking of my body was involuntary and out of my control. The brain told my body what to do and it saved my life, who knows.

What i should have told him was no….It wasn't me. I am thankful for the brain, your mind. It saved the day. The brain is so, so powerful.

The mind is truly a storehouse of thoughts, feelings and memories. I had forgotten about this incident that happened when i was twenty three or twenty four.

I had gotten a job with the city and a physical was required before starting. When i had the physical, a lump was found in one of my breast. I was scheduled to have a biopsy and on the eve of the biopsy while in the city hospital ward, a young woman asked me why was I in the hospital.

I told her that a lump was found in my breast. She loudly announced, don't go to sleep because if you do, you'll wake up with no breast. That's what happened to my friend.

With this knowledge, my mind was fixated on staying awake, don't go to sleep no matter what. I was recently married and i didn't want to go home to my husband with no breast.

While having the biopsy, I could hear everything. I heard someone say, go deeper and I could feel them going deeper and my mind was intent on staying awake until I knew they were finished cutting the tissue out and my breast was still there. My mind could not relax until this was over.

Thank god the result of the biopsy was benign.

How powerful is our mind, our brain that it would even override the medication.

The brain, your brain is powerful.

THE EYES

THE EYES ARE LIKE THE WINDOW TO SOMEONE'S SOUL. It's like piercing into the inner parts.

The eyes are beautiful with different shapes and various colors.

The eyes protect you by tearing up when irritants are present. It allows us to see others, to see our children, to see the animals, flowers, the outside world in all its various colors.

Not only does the eyes allow us to see, we can also decorate or enhance our eyelashes.

Let's have a little fun. Have you ever fixated your eyes on someone that you felt had mistreated you. Your eyes fixated on them as if staring at them is to punish them. I think we've all experienced this sometime in our lives. We use our eyes sometimes as if to freeze the person. How amazing.

The eyes are so powerful that you can actually feel when someone stares at you. The eyes are powerful.

 Can you see how awesome our bodies are. There is no other machine that can compare to your body

Take care of this organ because it will lead you into the light, where there is darkness.

I don't know all of the functions of the eyes , but there are many. I just want you to be aware of your body.

I'm just a layman saying how we need to know what we have in our bodies. We will never know everything about our bodies, but if we know what we have, we will be able to take care of this magnificent machine.

It is obvious that we are all made fearfully and wonderfully. The earliest that we know what we have, we will be able to make better choices in our lifestyle. It has been said, if you knew better, you would do better.

That applies here.

We can wink with our eyes; sometimes we scare our children with our eyes. We even flirt with our eyes..

All moms have done this when trying to get their children's attention when they do something displeasing. Mom will give you that certain glance, it's more like a deep heartfelt look that tells you exactly what she's saying. It says stop, now, or else. You get the message. She literally means to stop, now or else.

THE HEART

THE HEART IS THE TOUGHEST ORGAN IN THE BODY.

The heart has an awesome task. It is working all the time, 24/7, 365 days transporting blood throughout the body. While we sleep, the heart is pumping blood, it really never sleeps.

The heart is the lifeline.

This organ has many muscles and is responsible to carry, again the blood throughout the body. Heart failure can happen when the heart can no longer pump blood efficiently to the body tissues and the lungs.

Your heart weighs less than a pound. That's amazing for all that the heart does, pumping blood all day.

The heart has two pumps. One pump sends blood to the lungs and the other pump sends blood around the body.

You can see how important the heart is and how to keep our blood pressure at a rate that is healthy.

When you go to the doctor, they always check your blood pressure. Your blood pressure can change. It can change if you're ill, stressed, worried, nervous or frightened.

The heart is one of those organs that can be transplanted to save another human beings life. If successful, the recipient could receive a young, healthy heart. That is truly awesome and amazing.

You can open up your heart to new experiences. Your emotions run deeply with this organ.

You feel deeply and love greatly.

Take care of this organ. Talk to your parents, your grandparents, aunts and uncles about the health history within your family.

By doing this, you will see what you are predestined to the heart

Possibly get, and you will be able to take the necessary precautions.

By knowing early what you possibly may get, you can see what you need to do to minimize the outcome. You will be able to make an informed decision on your lifestyle.

Your body is truly amazing.

Don't take this organ for granted for it has an awesome job to perform. Your body let's you know what's going on within.

If you feel tired constantly, don't ignore it. It's telling you to seek medical help. Sometimes we ignore the signs, but if we take heed, you may prevent a disaster later on in your health.

Your body does alert us if something is wrong in our body. We must recognize what it is telling us. We have been given a precious gift.

I'm speaking from experience. Had I recognized that the tired feeling that I was having was that something was wrong in my body, I would have avoided a heart attack. Thank god the heart attack wasn't fatal and now I pay attention to my body.

In fact, my heart attack occurred when I was on my way to a treatment center.

What an experience.

I was driving while having a heart attack. God was with me because I could have killed the people on the road.

I recall while driving, a lot of car horns beeping and little did i know that they were honking at me. Apparently I was driving reckless. Thank god was protecting others and me.

When I got to the treatment center I parked the car and a police officer asked me where was I going; I said here and he said no, I'm going to call an ambulance for you because you need to go to the hospital. Immediately the ambulance was there and I remember being placed in the ambulance and rushed in the hospital where i found myself being administered to right away. And after a while a flood of doctors was asking me a lot of questions.

The food that was served was like medicine. I didn't realize at the time, but the heart attack took my taste away.

I had to go to therapy after the heart attack and my taste finally came back. After reflecting on what happened during my heart attack, I wanted to thank the police officer for stopping me from going to the treatment center and for calling the ambulance for me.

I never saw a police officer at that site ever again. I honestly believe the police officer was an angel. You see, if i had gone to the center i would have died because i didn't have an appointment and would have died waiting to be seen. By the police officer intervening and getting me to the hospital, he saved my life.

The curious thing is that the hospital was across the street from the center. You never know who you're talking to. The stranger can be an angel in disguise, sent by god,

I am very much aware of my body and I implore you to pay attention to it. It can save your life.

You are truly fearfully and wonderfully made.

THE THROAT-MOUTH

THE THROAT GIVES US OUR UNIQUE VOICE. Every one's voice is definitely different. No one sounds like you. You can be imitated, but not duplicated.

You are unique. First, we can speak, our form of communication, we can sing, whistle and so much more.

After the mouth, the throat and esophagus, the next major section of the digestive tract are the stomach and the small intestine.

The gastric juices in the stomach helps us in the digestive process.

Sometimes our stomach talk to us by gurgling or making noises. The stomach houses lots of nutrients necessary for our existence.

Some people refer to the stomach, as the gut.

Some people may have a lean looking stomach and some people have a stomach that is protruding. The protrusion may indicate that there may be a problem in their health. That's the time to take heed to what your body is trying to tell you. It's time to check it out.. Go to your doctor. .

The tongue within the mouth is the body's most flexible muscle.

Saliva in the mouth is composed of 99.5 Percent water and it lubricates our food to make swallowing and chewing easier.

The throat houses other organs like the esophagus. This organ is multi functional and unique.

The throat is the organ of speech.

The human body is so complex and each organ performs a single task or a list of numerous functions.

Your body multitask.

The spectacular human body.

God loves his children that he gave them everything they would need.

Our organs perform different functions. They multitask. No other machine can ever compare.

Your machine truly is like no other.

THE TEETH

OUR TEETH HELP US TO CHEW OUR FOOD. It breaks down the food for us and it helps us to speak correctly.

You can brush your teeth. Your teeth, if broken or a cavity needs to be filled, can be treated by going to your dentist. You need to keep up the routine of brushing and flossing your teeth daily.

The teeth can be repaired. You can get implants, crowns, dentures and very popular with the youth are veneers. .

One place on your face that people look at are your teeth, your mouth. Without your teeth your appearance would not be attractive. It will give your face a different look and having teeth make you look healthy.

Your smile is beautiful and infectious when you have healthy, clean looking pearly teeth.

I have seen some unattractive people with unattractive teeth, but when they got healthy looking teeth, they became very attractive. It's like the teeth framed, polished their whole face.

A bright, clean set of teeth, glistening when you smile is so beautiful. So please take care of them. Eat properly and brush and floss often and see your dentist regularly.

THE EARS

THE EARS ARE CRUCIAL. All of your organs are important. God blessed you with a body that can hear others as well as hear even a whisper. You can hear external noises, such as a bird chirping, children playing, you can hear crying, laughter and even someone calling out for help.

This is an organ that you can embellish by wearing earrings or several earrings. You can truly beautify this organ

This organ is essential for keeping you balanced.

It is wonderful to be able to hear the wonderful sounds in this beautiful world.

Intense or extreme sounds can damage the ears. Upon hearing extremely loud noises, the ears will react. You can damage your hearing.

We are able to hear music and the beautiful songs, fireworks and the laughter of our loved ones.

Pamper your ears. Take care of this organ. It is spectacular. All of our organs are.

You can become dizzy or have an uncomfortable feeling if you stand up too fast or if you change heights rapidly.

Take care of this organ. You need it.

You are truly fearfully and wonderfully made. Marvelous are thy works.

Psalms 139-14.

THE ARM

THE ARM CONSISTS OF THE ARM, the forearm, the wrist connected by the hand. This vital organ helps us to hold or to pick up things.

We can hug our parents, our children, our family, friends and strangers.

We all get vaccinated with various shots, such as immunization shots for children and other shots we must get are flu shots. There are shots for all types of diseases.

Invariably when we get these shots, they are administered in our upper arm. I believe it is because this area of the arm is where the shot would be most effective.

When we go for our physical, the clinician will take your blood pressure by hooking up your arm to a machine.

This organ, also can be embellished. We wear bracelets, either on the upper arm or the lower arm. Your arms do so much more. They are beautiful.

THE FACE

THE FACE COMES IN DIFFERENT SHAPES AND SIZES.

The shapes are round, diamond, rectangle and oval. All are beautiful. Do you know your shape.

The face can have interesting features, such as moles or dimples. Some people say these are freaks of nature. If this is so, let the freak of nature continue because these features are lovely and interesting..

The eyebrows on the face really frames the face. Some eyebrows are so beautiful. Some eyebrows are full and some are sparse, I admire the different shapes,

FINGERS-FINGERNAILS

FINGERNAILS

Fingernails helps us to pick up things also. Nails can be filed into different shapes, some pointed, some square.

You can decorate them by painting them. You can put jewels on them. Fingernails allow many people to earn a living by being a manicurist or a nail tech.

How awesome is our god that he gave us function, beauty and a means to make money, yes, a living.

In addition to painting the fingernails and toenails, god allowed us to grace our nails with beautiful stones he left in the earth to be polished and ready to wear. The most familiar stones are; diamonds, emeralds, sapphire and ruby.

There are many more stones that are beautiful and valuable and very popular. Some of them are; chrome dioxide, black spinel and tanzanite.. God is so mindful and concerned about our needs.

Tanzanite is a relatively new found stone as opposed to the diamonds, rubies, etc.

Tanzanite comes in different hues' it is a bluish purplish color. All shades are lovely, but the darker the color, the more valuable it is.

So if you have any tanzanite, hold on to it, because in time it will become more valuable.

You or your children may be the one to discover a valuable stone. The earth hasn't been totally discovered yet. Who knows.

God gave us the ability to make rings that we can wear. On our hands. Fingers, fingernails are functional as well as beautiful.

What an organ.

See how functional your body is as well as being beautiful.

You are thoughtfully and wonderfully made.

THE FEET

THE FEET HELPS US TO STAND, to walk, to run, skip and so much more. We depend on our feet to support us.

Our feet have the most bones in our body.

You can also paint your toenails. We are blessed to have an organ that functions as well as one that we can beautify.

This organ has a tremendous task to perform. The feet needs to be protected so it can serve us well.

We sometimes put a lot of pressure on our feet. We may overwork them with no relief in mind. Gaining weight can cause our feet to work harder.

Soak your feet. Give them some relief. They in kind will take care of you.

This organ works so hard to perform and to carry us throughout the day.

We should pamper, pamper this organ. We rely on our feet.

When there is no vehicle in sight to take us places, the feet always carry us.

People have been walking for years. It is reliable. Appreciate them.

Your feet supports you.

REPRODUCTIVE SYSTEM

THE FEMALE AND MALE ORGANS CAN PRODUCE A HUMAN BEING.
What other machine can do that.

The male organ produces the sperm, the seed to fertilize the female egg, resulting in the pregnancy where life begins. This truly is a miracle.

The pregnant female's body will go through a lot of bodily changes preparing for the birth of the child.

The female is most likely to gain weight because she is not only eating for herself, but for the unborn child as well.

God said you are fearfully and wonderfully made. When the child is born, the physician can tell what gender the child is by his/her genitals.

It is wonderful to see the development of a baby. To see it grow from an infant to an adult.

We often say, how you have grown after seeing someone we hadn't seen for a while.

Did the person's skin stretch or did their body become elongated.

That is a miracle to me because you don't actually see them grow.

Your body is a miracle.

THE SKIN

THE SKIN IS THE LARGEST ORGAN IN THE BODY.

The skin suffers more physical damage than any other body organ because of its location.

The skin has different layers. The outer layer is a protecting covering and the underlying skin is made up of many different tissues doing various functions.

The outer skin is so special in protecting us and it can renew itself.

The skin is our receptor. If you put your hand on something hot, the skin feels it and it responds with a shout or cry out.

Don't expose your skin too much in the sun because it can cause harm and too much sun can cause skin cancer.

The skin reacts to sunlight. In biology, there is no such thing as race. Our body produces a molecule called melanin. It colors the skin, our hair.

God is no respect of person that he made us all the same. No skin tone is greater than the other. We are all equal.

We all have a brain, two eyes, two arms, legs, feet. You get the message. We all equally have to take care of our body.

We all need to use sunscreen, regardless of our skin color. Your skin color can also be determined by your location. There are people in africa that are blonde with hazel eyes.

THE NOSE

outside. This organ is external, you can see it.

When you inhale and exhale you can sometimes feel your nostril movement. The nose has hair inside to prevent irritants from invading too far inside, a protective measure. God thought of you.

One of the functions of the nose is to smell, and that it does.

The nose let's you enjoy the sweet fragrance of flowers, the smell of others.

The smell of food, so savory is one of the sweetest joys we can and do have with this organ.

The nose can also alert you against a material that may be harmful to you, such as ammonia if you inhale it.

This organ is amazing.

THE BONES

THE BONES ARE COMPRISED OF CONNECTING TISSUE that is very strong like steel, but is very light.

Your bones can constantly break down, but your bones can rebuild itself during the growing process. The bone adjust its size and shape.

What a highly complex machine we have in our body. See what a wonderful body god blessed us with. He gave us everything we need to sustain us. How thoughtful and precise is our creator.

We can prolong our life if we take care of our magnificent body.

MORE FACTS ABOUT
THE HUMAN BODY

Our body consists of many fluids. Some are;

- Water

- Blood

- Saliva

- Mucous

- Tears

- Gastric juices

We are truly blessed.

This book is an easy read and easy to understand. All ages can benefit from it if you take heed.

Professionally, I know nothing about the human body, but wha tI do know is that god provided us with a body that would perform and sustain us for the remainder of our days.

I now realize that god wants us to be selfish. That means taking care of yourself first.

That is self preservation.

After taking care of yourself first, then you can help others. A lesson learned.

It's ok to be selfish.

There are so many more parts of the body that was not mentioned in this book. I briefly talked about a few, because i wanted to get your attention to acknowledge your body and what it was designed to do.

Your body is meant to take care of you in the most efficient way.

In knowing this, it will equip you to not take it lightly for it has an awesome job.

I hope you look at your body in a different light. You can make it if you try. Just be mindful of what you have.

I will praise thee for I am truly blessed.

THE HAIR

YOUR HAIR IS YOUR GLORY. It warms your head. It has been said that if your head and feet are covered, you can be warm.

Hair is lustrous and beautiful. Hair comes in different colors. There is black hair, brown hair, red hair, blonde hair, gray hair and white hair. You can also color your hair any color you desire.

The hair can be cut and it grows back. You can wash your hair, dry your hair, even relax your hair and style your hair, the choices are limitless.

Some people like to upkeep their hair themself and others frequent the beauty shop. The males also frequent their barber on a regular basis to stay on top of the most recent trends.

The barber shop is also a lucrative business. It is so much more than it use to be. You can get a haircut, a fade, lineup, beard trimmed or shaved.

See how versatile our bodies are. It is both functional, practical and beautiful.

That is why we need to see our body as a blessing.

God is highly intelligent that he made us with a body that would sustain us throughout our life.

We have the most sophisticated, complex body. I hope you realize how fearfully and wonderfully made you are.

God is the only one that is truth, he is the light and He is love.;

TALKING ABOUT THE HUMAN BODY

GOD IS SO SKILLFUL that when different organs function in the body, we don't feel virtually anything and it doesn't hurt. That's amazing.

In my estimation, god love colors. You can see it in the earth, the trees, the flowers, all in different colors. The animals are so beautiful in different shapes and sizes.

God made everything and was pleased, but when he made mankind, that was his greatest creation .

Even the dirt is colorful. There is dirt that is red. The human body

The jewels also are in many different hues. The eyes, the hair and the planets have different colors.

God showed his love of colors when he created mankind, the human body.

We live in a colorful world and god intended for us to enjoy it.

Think of it this way, the variety of colors make life more joyful and fun and certainly never boring.

God is so clever and funny. He definitely have a personality.

I am so inspired how intelligent our Creator is. He is so thoughtful and mindful of every detail in our body.

God is so smart when he created heaven and earth. God just spoke the word and it became a reality.

When god made us, there was no lack. Our body is 100 the human body

Percent suitable, so adequate.

You are beautiful.

I would like to reiterate that the purpose of this book is to enlighten you in my humble opinion about your beautiful body.

Love your body. It is worth your attention to pamper your magnificent body. Take care of it and it will take care of you, if you let it.

God is so marvelous in his works.

This book is an easy read with the large print, easy to see. This book hopefully will be easy to understand the importance of taking care of your body.

I will praise thee for I am fearfully and wonderfully made.
Marvelous are thy works and my soul knoweth right well.

Psalms 139;14..

Thanks be to god, our creator, for guiding me in this endeavor. This book.

God the father, god the son, and god the holy ghost.

I want you to start journaling. Each day start with thanking god for a new day, date each day

Write down what you hope to achieve that day. I have provided you with empty pages in this book to start your new journey. If you have nothing to record, just thank god for the day. You can always look in your journal to see what has transpired. I think you will be amazed.

This book is in memory of my beloved parents who loved me unconditionally and taught me about the love of God, the love of family, friends, and strangers.